MY FIRST Counting with Mermaids BOOK

Jadedra Gilmore-Barber

My First Counting with Mermaids Book © 2024 Jadedra Gilmore-Barber

All rights reserved. This book or any portion thereof may not be reproduced or used in any manner whatsoever without the express written permission of the publisher except for the use of brief quotations in a book review.

Published by: EarKanDee LLC
For more information: earkandee.educate@gmail.com
ISBN: 979-8-9915315-4-2
www.earkandeeonline.com

"If better is possible, good is not enough."
— Benjamin Franklin

TABLE OF CONTENTS

Count Forwards..1
zero...2
one..3
two..4
three...5
four...6
five..7
six...8
seven..9
eight..10
nine...11
ten...12
Count Backwards..13
ten...14
nine...15
eight..16
seven..17
six...18
five..19
four...20
three...21
two..22
one..23
zero...24
SPLASH..25

LET'S COUNT FORWARDS!

zero

I see 0 objects.

one

I see <u>1</u> treasure chest.

I see <u>2</u> sea stars.

three

I see <u>3</u> seahorses.

four

I see <u>4</u> sea turtles.

five

I see 5 sting rays.

seven

I see 7 seashells.

eight

I see <u>8</u> octopuses.

nine

I see 9 sea urchins.

ten

10

I see <u>10</u> whales.

12

LET'S COUNT BACKWARDS!

ten

10

There are <u>ten</u> whales.

nine

There are <u>nine</u> sea urchins.

eight

There are <u>eight</u> octopuses.

seven

7

There are <u>seven</u> seashells.

five

There are <u>five</u> sting rays.

19

four

There are <u>four</u> sea turtles.

three

There are <u>three</u> seahorses.

There are <u>two</u> sea stars.

one

There is <u>one</u> treasure chest.

zero

There are <u>zero</u> objects.

SPLASH!

Biography

Dr. Jadedra Gilmore-Barber was born and raised in Fort Valley, Georgia. She has been an educator for over 14 years. Dr. Gilmore-Barber has held various roles during her years in the field of education. She has served as a teacher, academic coach, and administrator. She graduated from Georgia Southwestern State University with a bachelor's degree in Early Childhood Education, a master's degree in Curriculum and Instruction, and a specialist degree in Teacher Leadership. She graduated from Columbus State University with a specialist add-on in Educational Leadership, and a doctoral in Educational Leadership. Dr. Gilmore-Barber's philosophy of education is "Where better is possible, good is not enough!" She strives to motivate students to gain and maintain a love of learning.

EARKANDEE LLC

LISTEN | DESIGN | DISCOVER

"Where better is possible, good is not enough."

PUBLISHING "EFFECTIVE"
EDUCATIONAL RESOURCES GLOBALLY

AVAILABLE NOW ON AMAZON

"MUST LEARN SIGHTWORDS" THROUGH READING, WRITING, AND MATH WORKBOOK (KINDERGARTEN) *AVAILABLE NOW*	"MUST LEARN SIGHTWORDS" THROUGH READING, WRITING, AND MATH WORKBOOK (FIRST GRADE) *AVAILABLE NOW*	"MUST LEARN SIGHTWORDS" THROUGH READING, WRITING, AND MATH WORKBOOK (SECOND GRADE) *AVAILABLE NOW*	MY FIRST COUNTING IN SPACE BOOK *AVAILABLE NOW*
MATH "GROWTH" JOURNAL GRADE LEVELS K-5 *AVAILABLE NOW*	MATH "GROWTH" JOURNAL GRADE LEVELS 6-8 *AVAILABLE NOW*	MATH "GROWTH" JOURNAL GRADE LEVELS 9-12 *AVAILABLE NOW*	MY FIRST COUNTING IN SPORTS BOOK *AVAILABLE NOW*
WRITING ON THE ROADS OF THE USA "STATES": K-2 HANDWRITING WORKBOOK *AVAILABLE NOW*	G.R.E.A.T. IS FOR A DAY WITH MY GRANDPARENTS! CHILDREN'S BOOK *AVAILABLE NOW*	THE MIGHTY MOVE UPSTAIRS CHILDREN'S BOOK *AVAILABLE NOW*	MY FIRST COUNTING IN THE SNOW BOOK *AVAILABLE NOW*

Stay Tuned For More....

DON'T FORGET TO REVIEW THIS WORKBOOK ON AMAZON

Find us at:
www.earkandeeonline.com

Contact us:
earkandee.educate@gmail.com

www.ingramcontent.com/pod-product-compliance
Lightning Source LLC
Chambersburg PA
CBHW040029050426
42453CB00002B/56